Mobile Apps for Kids:

Disclosures Still Not Making the Grade

Contents

FTC Staff Report[1]

Introduction

In February 2012, Federal Trade Commission (FTC) staff issued a report on a survey of mobile "apps" offered for children in Apple's App Store and Google's Android Market, the two largest U.S. app stores. The report, *Mobile Apps for Kids: Current Privacy Disclosures are Disappointing*,[2] found that little or no information was available to parents about the privacy practices and interactive features of the mobile apps surveyed prior to download. As a result, the report called on all members of the kids' app ecosystem – app stores, developers, and third parties that interact with the apps – to provide greater transparency about the data practices and interactive features of apps geared to children. The report stated that FTC staff would conduct a follow-up survey in six months to evaluate whether and how industry had addressed the concerns raised.[3]

FTC staff conducted its follow-up survey during the summer of 2012. Like the first survey, the new survey examined the disclosures that apps provided about their privacy practices and interactive features, such as links to social media. However, the new survey went a step further by testing the apps' practices and comparing them to the disclosures made. Specifically, the new survey examined whether the apps included interactive features or shared kids' information with third parties without disclosing these facts to parents. The answer: Yes, many apps included interactive features or shared kids' information with third parties without disclosing these practices to parents.

---

1. The primary authors of this FTC staff survey and report are Manas Mohapatra and Andrew Hasty of the FTC's Mobile Technology Unit. They received valuable assistance from FTC summer law clerks Amy Greenspan, Kristen Poppenhouse, and Batool Raza, and staff from throughout the Bureau of Consumer Protection. Ryan Sandler of the Bureau of Economics provided valuable assistance reviewing the survey results, and Jessica Skretch of the Division of Consumer and Business Education created the graphics, charts, and design of the report.

2. The first report, hereafter referred to as "Mobile Apps for Kids Report," is *available at* http://www.ftc.gov/os/2012/02/120216mobile_apps_kids.pdf.

3. The recommendations in this report and the prior report are designed to encourage "best practices" by companies in the kids' app ecosystem. Staff did not examine whether the practices observed violated the laws enforced by the Commission, and some of staff's recommendations may go beyond what would be required to comply with the law.

Since issuing the first kids' app report, the Commission has continued to promote consumer protections in mobile technologies by engaging in a host of policy, enforcement, and educational initiatives.[4] Additionally, other government agencies, including the California Attorney General and the U.S. Department of Commerce, have launched efforts to increase transparency in the mobile marketplace,[5] and several trade associations have issued self-regulatory guidelines or launched initiatives regarding mobile app privacy and related issues.[6] Likewise, Apple and Google recently announced changes to their app stores that may address

---

4. In March 2012, the Commission issued a final Privacy Report, which set forth best practices for businesses to protect consumers' privacy and give them greater control over the collection and use of their personal data, including when using mobile devices. FTC Report, *Protecting Consumer Privacy in an Era of Rapid Change* (March 26, 2012) ("Privacy Report"), *available at* http://www.ftc.gov/os/2012/03/120326privacyreport.pdf. Commissioner Rosch dissented from the issuance of the Final Privacy Report. *See id.* at Appendix C. Further, in May 2012, the Commission held a workshop to bring together representatives from industry, academia, and consumer organizations to discuss how businesses can make effective disclosures in new media, including on mobile devices. FTC Workshop, *In Short: Advertising & Privacy Disclosures in a Digital World* (May 30, 2012), *available at* http://www.ftc.gov/bcp/workshops/inshort/index.shtml. FTC staff also published guidelines to assist mobile app developers observe and comply with truth-in-advertising and privacy principles. *See* FTC, *Marketing Your Mobile App: Get It Right From the Start* (August 2012), *available at* http://business.ftc.gov/documents/bus81-marketing-your-mobile-app. In addition, the Commission has proposed modifications to the Commission's Children's Online Privacy Protection Rule, in part to clarify the consumer protections that should apply when children use mobile devices. Press Release, FTC, *FTC Seeks Comments on Additional Proposed Revisions to Children's Online Privacy Protection Rule* (Aug. 1, 2012), *available at* http://www.ftc.gov/opa/2012/08/coppa.shtm.

5. In February 2012, the California Attorney General announced an agreement with the six leading mobile app platforms that was designed to ensure that mobile apps available through those platforms post privacy policies for consumers to view. Press Release, State of California Department of Justice, *Attorney General Kamala D. Harris Secures Global Agreement to Strengthen Privacy Protections for Users of Mobile Applications* (Feb. 22, 2012), *available at* http://oag.ca.gov/news/press-releases/attorney-general-kamala-d-harris-secures-global-agreement-strengthen-privacy. In June 2012, the U.S. Department of Commerce announced that it was convening a privacy multi-stakeholder process to address mobile app transparency, with the goal to have stakeholders develop voluntary, enforceable codes of conduct. Press Release, National Telecommunications & Information Administration, Department of Commerce, *First Privacy Multistakeholder Meeting: July 12, 2012* (June 15, 2012), *available at* http://www.ntia.doc.gov/other-publication/2012/first-privacy-multistakeholder-meeting-july-12-2012.

6. *See, e.g.*, Press Release, GSMA, *GSMA Announces New Initiative Addressing Mobile App Privacy* (Feb. 27, 2012), *available at* http://www.gsma.com/newsroom/gsma-announces-new-initiative-addressing-mobile-app-privacy/; Press Release, ACT, *ACT Introduces the App Privacy Icons* (Oct. 4, 2012), *available at* http://actonline.org/act-blog/archives/2674; CTIA, *Best Practices and Guidelines for Location Based Services*, *available at* http://www.ctia.org/consumer_info/service/index.cfm/AID/11300; Application Developers Alliance, *Privacy Summit Series*, *available at* http://devprivacysummit.com/.

concerns about the failure to post privacy policies and the transmission of personal information by apps.[7]

These efforts have the potential to improve the information available to parents about the apps their kids use. Since the first kids' app report was issued, the market for mobile apps has continued to grow at an explosive rate, providing many benefits and conveniences to consumers. As of September 2012, there were over 700,000 apps available in Apple's App Store, a 40% increase since December 2011, and over 700,000 apps available in Google Play,[8] an 80% increase since the beginning of 2012.[9] The rise in the number of apps corresponds to the increasing number of U.S. adults who own devices capable of using apps. According to the Pew Research Center, nearly nine out of ten U.S. adults have a cell phone and more than 40% of these cell phone owners download apps to their phones.[10]

As consumers' embrace of the mobile marketplace has increased, so have their concerns about what mobile apps do with their personal information. For example, a recent Pew study found that 54% of app users decided not to install an app once they discovered how much

---

7. In August 2012, Google updated its developer program policy to state that "apps that disclose personal information without authorization are not allowed." Ingrid Lunden, *Google Tightens up App Policy, Gets Stricter on Naming/Icon, Payments, Privacy, Ads and Spam Rules [Developer Letter]* (Aug. 1, 2012), *available at* http://techcrunch.com/2012/08/01/google-tightens-up-app-policy-gets-stricter-on-namingicon-payments-privacy-ads-and-spam-rules-developer-letter/. In September 2012, Apple released the newest version of its iOS operating system. The new version allows developers to insert a link to a privacy policy directly on the app promotion page. *See Adding New Apps*, iTunes Connect Developer Guide, *available at* http://developer.apple.com/library/mac/#documentation/LanguagesUtilities/Conceptual/iTunesConnect_Guide/8_AddingNewApps/AddingNewApps.html#//apple_ref/doc/uid/TP40011225-CH13-SW1.

8. In March 2012, the Android Market became a component of Google Play. *See* Jamie Rosenberg, *Introducing Google Play: All your entertainment, anywhere you go*, Google Official Blog (Mar. 6, 2012), *available at* http://googleblog.blogspot.com/2012/03/introducing-google-play-all-your.html.

9. *See* Don Reisinger, *Can Apple's App Store maintain its lead over Google Play?*, CNET (Sept. 27, 2012), *available at* http://news.cnet.com/8301-1035_3-57521252-94/can-apples-app-store-maintain-its-lead-over-google-play/; Brian Womack, *Google Says 700,000 Applications Available for Android,* Bloomberg News (Oct. 29, 2012), *available at* http://www.businessweek.com/news/2012-10-29/google-says-700-000-applications-available-for-android-devices. In March 2012, when the prior Commission Staff report was issued, there were more than 500,000 apps in the Apple App Store and 380,000 apps in the Android Market. *See* Mobile Apps for Kids Report, *supra* note 2, at 1.

10. *See* Jan Lauren Boyles et al., *Privacy and Data Management on Mobile Devices* (Sept. 5, 2012) ("Pew Mobile Privacy Survey"), *available at* http://pewinternet.org/~/media/Files/Reports/2012/PIP_MobilePrivacyManagement.pdf. According to Pew, 45% of U.S. adults have a smartphone, up from 35% who owned a smartphone in May 2011, and 29% of U.S. adults own either a tablet computer or an e-reader, up from 18% in January 2012. *See* Joanna Brenner, *Pew Internet: Mobile*, Pew Internet & American Life Project (Sept. 14, 2012), *available at* http://pewinternet.org/Commentary/2012/February/Pew-Internet-Mobile.aspx. Of this group of app users, one in three has downloaded an app to their mobile device for use by a child. Amanda Lenhart, *Downloading Apps for Children* (May 15, 2012), *available at* http://pewinternet.org/Commentary/2012/May/Downloading-apps-for-children.aspx.

personal information the app would collect. The study also showed that 30% of app users have uninstalled an app that was already on their cell phone because they learned that the app was collecting personal information the users did not wish to share.[11] Consistent with these findings, a recent study by the Berkeley Center for Law and Technology showed that most consumers consider the information on their mobile devices to be private.[12]

Staff conducted its follow-up survey in the midst of these developments in the mobile marketplace. The survey results showed that parents still are not given basic information about the privacy practices and interactive features of mobile apps aimed at kids. Indeed, most apps failed to provide *any* information about the data collected through the app, let alone the type of data collected, the purpose of the collection, and who would obtain access to the data. Even more troubling, the results showed that many of the apps shared certain information – such as device ID, geolocation, or phone number – with third parties without disclosing that fact to parents. Further, a number of apps contained interactive features – such as advertising, the ability to make in-app purchases, and links to social media – without disclosing these features to parents prior to download.

Thus, despite many high-visibility efforts to increase transparency in the mobile marketplace, little or no progress has been made. As a result, Commission staff is taking additional steps to increase the focus on this important issue:

First, FTC staff strongly urges the mobile app industry to develop and implement "best practices" to protect privacy, including those recommended in the recent FTC Privacy Report: (1) incorporating privacy protections into the design of mobile products and services ("privacy by design"); (2) offering parents easy-to-understand choices about the data collection and sharing through kids' apps; and (3) providing greater transparency about how data is collected, used, and shared through kids' apps.[13] These standards should be developed expeditiously to ensure that consumers have confidence in the growing mobile apps marketplace.[14]

---

11. *See* Pew Mobile Privacy Survey, *supra* note 10, at 2.

12. The Berkeley Center survey reported that 78% of the U.S. consumers surveyed considered the information on their mobile phones at least as private as that on their home computers. The survey also found that 92% of respondents said that they would "definitely" or "probably" not allow the use of their locations to be used to tailor advertising for them. Jennifer M. Urban, et al., *Mobile Phones and Privacy* (July 11, 2012) at 9, 20, *available at* http://papers.ssrn.com/sol3/papers.cfm?abstract_id=2103405.

13. *See* Privacy Report, *supra* note 4 at vii-viii.

14. As noted above, the Department of Commerce is leading an effort to address one of these recommendations – mobile application transparency – through a multistakeholder process.

Second, FTC staff is developing and will soon issue consumer education directed to parents to help them navigate the mobile app marketplace and avoid apps that fail to provide adequate disclosures about how children's information will be used.

Third, FTC staff is launching multiple nonpublic investigations[15] to determine whether certain entities in the mobile app marketplace have violated the Children's Online Privacy Protection Act ("COPPA"), or engaged in unfair or deceptive trade practices in violation of the FTC Act.[16]

Fourth, FTC staff will conduct a third kids app survey once the initiatives and activities described in this report, including the Department of Commerce multistakeholder process and other self-regulatory efforts, have had a reasonable time to develop.

The survey results described in this report paint a disappointing picture of the privacy protections provided by apps for children. These findings should spur more resolute action by industry to address this important issue.

Survey Overview and Recommendations

FTC staff conducted its follow-up survey during the summer of 2012 to examine whether child-related apps were disclosing key information to parents prior to download. In addition to examining the apps' disclosure practices, staff downloaded and used the apps to learn about their data collection and sharing practices and to determine whether certain interactive features were present.

Like the first survey, the new survey selected the apps to review by searching the Apple and Google Play app stores using the keyword "kids," and collecting the app promotion pages for the first 480 results returned by each app store. Next, staff randomly selected 200 apps from each store and closely reviewed the apps' disclosures. Specifically, staff looked at the disclosures and links on each app's promotion page, on the app developer's website, and within the app itself. Staff downloaded and tested the 400 apps to determine whether they contained certain interactive features (advertising, the ability to make in-app purchases, and

---

15. FTC investigations are nonpublic and the Commission does not ordinarily publicly disclose the subjects of its investigations before it issues a complaint or a settlement.

16. As noted above, the purpose of staff's survey was to examine the disclosures and practices of kids' apps to determine whether parents can make informed decisions before downloading apps for their kids. Staff has not made a determination as to whether the disclosures and practices examined violated the specific provisions of COPPA or constituted unfair or deceptive practices under the FTC Act.

links to social media) and whether they collected or transmitted any information from the mobile devices they were tested on.

Overall, staff found that a majority of the apps surveyed collected or transmitted information from the mobile device. Indeed, nearly 60% (235) of the apps reviewed transmitted device ID to the developer or, more commonly, an advertising network, analytics company, or other third party.[17] And 14 of the apps that transmitted device ID also transmitted geolocation and/or phone number. By contrast, only 20% (81) of the apps reviewed disclosed any information about the app's privacy practices.

Staff also found a high incidence of interactive features within the apps that, in most cases, were not disclosed to users. Specifically, 58% (230) of the apps reviewed contained advertising within the app, while only 15% (59) indicated the presence of advertising prior to download. Further, 22% (88) of the apps reviewed contained links to social networking services, while only 9% (36) disclosed such linkage prior to download.

In addition, 17% (66) of the apps reviewed contained the ability to make purchases for virtual goods within the app, with prices for each purchase ranging from $0.99 in apps from both app stores, to $9.99 for Google Play apps and $29.99 for Apple store apps. Although both stores provided certain indicators when an app contains in-app purchasing capabilities, these indicators are not always prominent and, even if noticed, may be hard for many parents to understand.[18]

The results of the survey are disappointing. Industry appears to have made little or no progress in improving its disclosures since the first kids' app survey was conducted, and the new survey confirms that undisclosed sharing is occurring on a frequent basis. Staff did find a handful of app developers that were providing users with simple and short disclosures. However, such instances were far from the norm,[19] and most apps failed to provide basic information about what data would be collected from kids, how it would be used, and with whom it would be shared. It is clear that more needs to be done in order to provide parents with greater transparency in the mobile app marketplace.

---

17. The extent to which the transmission and collection of device ID raises privacy concerns depends in part on how it is used. See discussion below in the section *Device IDs: Why are they important?*, *infra* at 9.

18. See additional discussion of these indicators, *infra*, in the Survey Results section regarding In-App Purchases in Appendix II.

19. Further, some of these app developers engaged in practices that contradicted their "short and simple" disclosures. For example, a number of the apps that utilized icons to describe their practices wrongly stated that they did not contain advertising.

Survey Results

Privacy Disclosures and Practices of Surveyed Apps

Privacy Disclosures

In assessing the privacy disclosures of the 400 apps selected for review, staff measured the number of apps that contained or linked to a privacy policy or disclosure. Reviewers were asked to look for any disclosure on the app promotion page or developer website that was expressly titled "privacy," or any graphics or text that made obvious statements regarding information collection, sharing, or data practices. This expansive definition of a privacy disclosure was meant to capture user-friendly disclosures like those contained in icons, seals, or badges, in addition to official privacy policies and other terms of use agreements. Reviewers also downloaded and used each app to assess whether there were any privacy-related disclosures within the app itself.[20]

In the first kids' app survey, only 16% (64) of the apps reviewed provided parents with a link to a privacy policy or other disclosure *prior* to downloading an app.[21] When using this same methodology for the follow-up survey, staff obtained a similar result, finding that 20% (81) of the apps reviewed linked to general disclosure information, including a privacy policy.[22] In staff's view, information provided prior to download is most useful in parents' decision-making since, once an app is downloaded, the parent already may have paid for the app and the app already may be collecting and disclosing the child's information to third parties. Nevertheless, to gain a more complete understanding, the current survey also looked for privacy disclosures provided to parents after downloading an app. When these in-app disclosures were included in the calculation, the number of apps that contained privacy disclosures remained the same. Of the 400 apps reviewed, only 20% (81) contained any privacy-related disclosure on the app's promotion page, on the developer website, or within the app.

---

20. This would include privacy disclosures provided by the app developer during the use of the app, such as a link to the developer's privacy policy, or additional notification dialog boxes.

21. The disclosures other than a privacy policy included those that had labels such as "terms of use," "terms and conditions," "terms of service," "Legal Notices," and "disclaimers." *See* Mobile Apps for Kids Report, *supra* note 2, at 13.

22. Although a slightly larger proportion of apps reviewed in 2012 had general disclosures in this survey compared to the 2011 survey, the difference is not statistically significant.

The privacy disclosures that were provided also raised concerns. Many consisted of a link to a long, dense, and technical privacy policy that was filled with irrelevant information and would be difficult for most parents to read and understand.[23] Others lacked basic details, such as what specific information about a child would be collected, the reason for collecting such information, or what parties would obtain the information. For example, one app that shared device ID and geolocation with advertising networks had a misleading privacy disclosure that discussed features about the user interface of the app, but did not disclose the fact that advertising networks or analytics companies would be receiving information through the app:

> In order to keep the app 100% free, you will receive the following: Search shortcut icon on your home screen, Search shortcut on your bookmarks and browser homepage. This will help us bring you more cool apps like this in the future. You can delete the search shortcuts easily (Drag & Drop to the garbage), this will not affect the application in any way.

Another app that transmitted device ID, geolocation, and phone number to multiple advertising networks had a troubling privacy disclosure stating that the app does not share information with third parties:

> [App Name] may record user foot prints for the analysis purposes. Your email address and personal information are only stored if you subscribe to the newsletter or special offers. We do not share or sell this information to third parties in any sense except Government or state agencies for security purposes.

Information Collection and Sharing Practices

In addition to looking at the apps' disclosures, the current survey tested whether the apps reviewed actually shared information from children.

Staff tested the apps to determine whether specific information (a user's name, geolocation, birth date, email address, mailing address, phone number, or various device identifiers) was collected by app developers, advertising networks, analytics companies, or

---

23. One app's privacy policy consisted of language from an employment contract that would not be relevant to consumers: "This provision covers SUPPLIER's Privacy Policy and the methods SUPPLIER uses to safeguard CLIENT secrets. The parties agree and understand that CLIENT will entrust many business, marketing, technological, programming, modifications and any other trade or process secrets (hereinafter referred to as "Confidential Information") to Employee." A number of stakeholders have noted that the privacy policies of many mobile app developers are thousands of words long, and serve more as a disclaimer rather than offering consumers effective notice. *See* Kevin J. O'Brien, *Data-Gathering via Apps Presents a Gray Legal Area.*, NY Times (Oct. 28, 2012), *available at* http://www.nytimes.com/2012/10/29/technology/mobile-apps-have-a-ravenous-ability-to-collect-personal-data.html.

other third-parties. In total, staff found that 59% (235) of the 400 apps transmitted some information from a user's mobile device back to the developer or to a third-party.

The most common piece of information that was collected and shared was a user's device ID, a string of letters or numbers that uniquely identifies each mobile device. Indeed, all 59% of the apps that transmitted some information transmitted a device ID to the developer or, much more commonly, to a third-party. Of all the apps reviewed, 5% (20) transmitted the device ID back to the developer, while 56% (223) transmitted the device ID to ad networks, analytics companies, or other third parties.[24]

---

24. Of the apps reviewed, some transmitted information only to the developer, some transmitted information to both the developer and a third-party, and many transmitted information solely to a third-party.

Device IDs: Why are they important?

Device IDs are short strings of letters and/or numbers that uniquely identify specific mobile devices. Today's smartphones typically have multiple device IDs, each used for a different purpose. Some device IDs are used to enable services like Wi-Fi and Bluetooth, or to uniquely identify specific devices operating on the carriers' networks. Other device IDs, like Apple's "UDID" or Android's "Android ID," are used by apps, developers, and other companies to identify, track, and analyze devices and their users across various mobile services.

Companies can receive a wide variety of information about users through mobile apps, including data about the device (like a user's device model, carrier, operating system version, and language settings) and personal data (like a user's name, phone number, email address, friends list, and geolocation). If this information is collected with a unique device ID, it can be associated with previously collected data with the same unique device ID.

The extent to which the collection of device IDs raises privacy concerns depends in part on how it is used. Because device IDs are difficult or impossible to change, they can be used by apps, developers, and other companies to compile rich sets of data or "profiles" about individuals. However, the use of device IDs when necessary for specific internal operations, such as protecting against fraud and theft, site maintenance, maintaining user preferences, or authenticating users, would not raise the same concerns.

Concerns about the creation of detailed profiles based on device IDs become especially important where, as staff found, a small number of companies (like ad networks and analytics providers) collect device IDs and other user information through a vast network of mobile apps. This practice can allow information gleaned about a user through one app to be linked to information gleaned about the same user through other apps.

When an app transmits a device ID, it often sends additional information along with that identifier. Information that is sent along with the device ID may include the following usage data:

Data Point	Android Example	iOS Example
Device ID	samplecf750cfa0aabd35c4040ca983d	samplede1e151f303ef269e8dd9423 4e4db710ad2
App Name	SampleAppForKids!	SampleAppForKids!
App Version #	1.0.1	1.0.2
Developer	Sample Developer Co.	Sample Developer Co.
Time Stamp	7/11/2012, at 9:50:28 am EST	7/13/2012, at 11:36:14 am EST
Operating System	Android 2.3.7	iOS 5.1.1
Device Model	LG Viper 4G LTE	iPhone 4
Language Configuration	English, United States	Español, España
Carrier	Sprint	AT&T

Staff found a much lower prevalence in transmission of information other than device IDs and usage data. Staff observed that 3% (12) of the apps transmitted a user's geolocation and 1% (3) transmitted a device's phone number. Although these figures appear low, they raise concerns for several reasons. First, in every instance where an app transmitted geolocation or phone number, it also transmitted the user's device ID. As a result, the third parties that received this geolocation data or phone number could potentially add it to any data previously collected through other apps running on the same device.[25] Second, the information was often transmitted to advertising networks, with no disclosures regarding why the advertising

---

25. For example, in the current survey, one ad network received information from 31 different apps. Two of these apps transmitted geolocation to the ad network along with a device identifier, and the other 29 apps transmitted other data (such as app name, device configuration details, and the time and duration of use) in conjunction with a device ID. The ad network could thus link the geolocation information obtained through the two apps to all the other data collected through the other 29 apps by matching the unique, persistent device ID. *See* David Norris, *Cracking the Cookie Conundrum with Device ID*, AdMonsters (Feb. 14, 2012), *available at* http://www.admonsters.com/blog/cracking-cookie-conundrum-device-id ("Device ID technology is the ideal solution to the problem of remembering what a user has seen and what actions he or she has taken: over time, between devices and across domains. . . . Device ID can also help businesses understand visitor behavior across devices belonging to the same person or the same residence."); Jennifer Valentino-DeVries, *Privacy Risk Found on Cellphone Games*, Digits Blog, Wall St. J. (Sept. 19, 2011), *available at* http://blogs.wsj.com/digits/2011/09/19/privacy-risk-found-on-cellphone-games/ (noting how app developers and mobile ad networks often use device IDs to keep track of user accounts and store them along with more sensitive information like name, location, e-mail address or social-networking data).

networks needed it or how they would use it. Third, the apps responsible for the collection appear, from their app promotion pages, to have been downloaded hundreds of thousands of times, meaning that a significant number of consumers' geolocation information had been shared through this small group of apps. Finally, staff's findings likely underrepresent the true level of data collection, since methodological constraints prevented staff from fully replicating a real-life situation.[26]

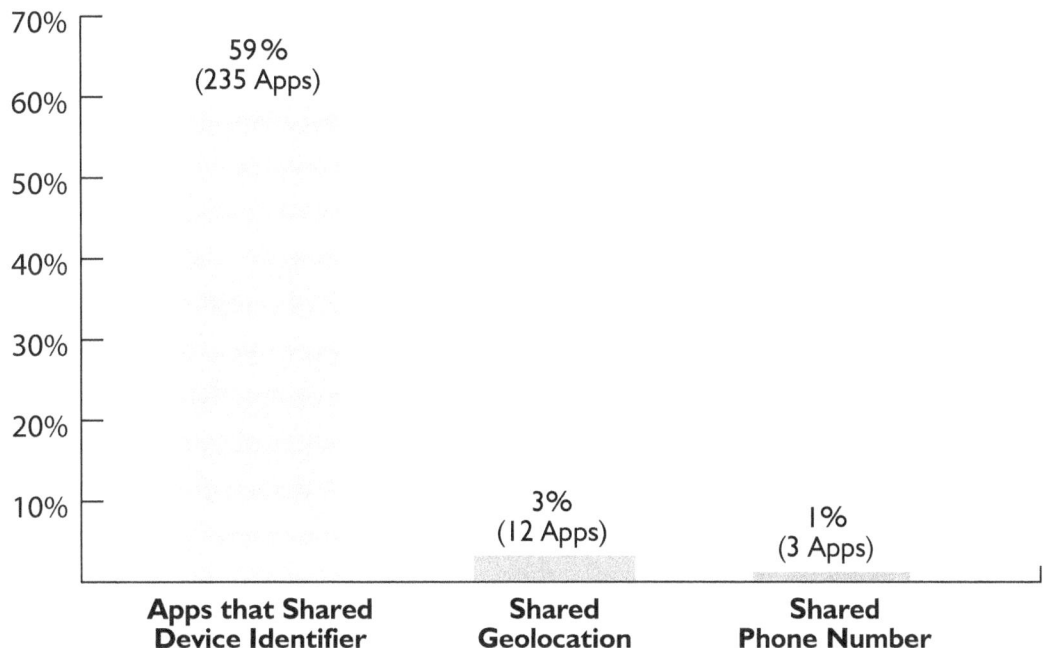

---

26. As described in greater detail in the methodology section, staff tested each app one time, in the same location, and on only one of a limited number of devices that were connected to the internet solely over Wi-Fi. To the extent that data transmission would be prompted by changes in a device's location, connection to a cellular tower, or other triggers, such data collection would not be captured or measured by staff's survey.

Shared Data Recipients

Each circle = **a company that received data** from at least 1 of the 400 apps we observed.

Size is based on the number of apps that the company received data from.

So, for example, the company that received data from 100 different apps is the largest and it's getting data from 25% of all the apps we observed.

Ad Networks, Analytics Companies, or other Third Parties

App Developers

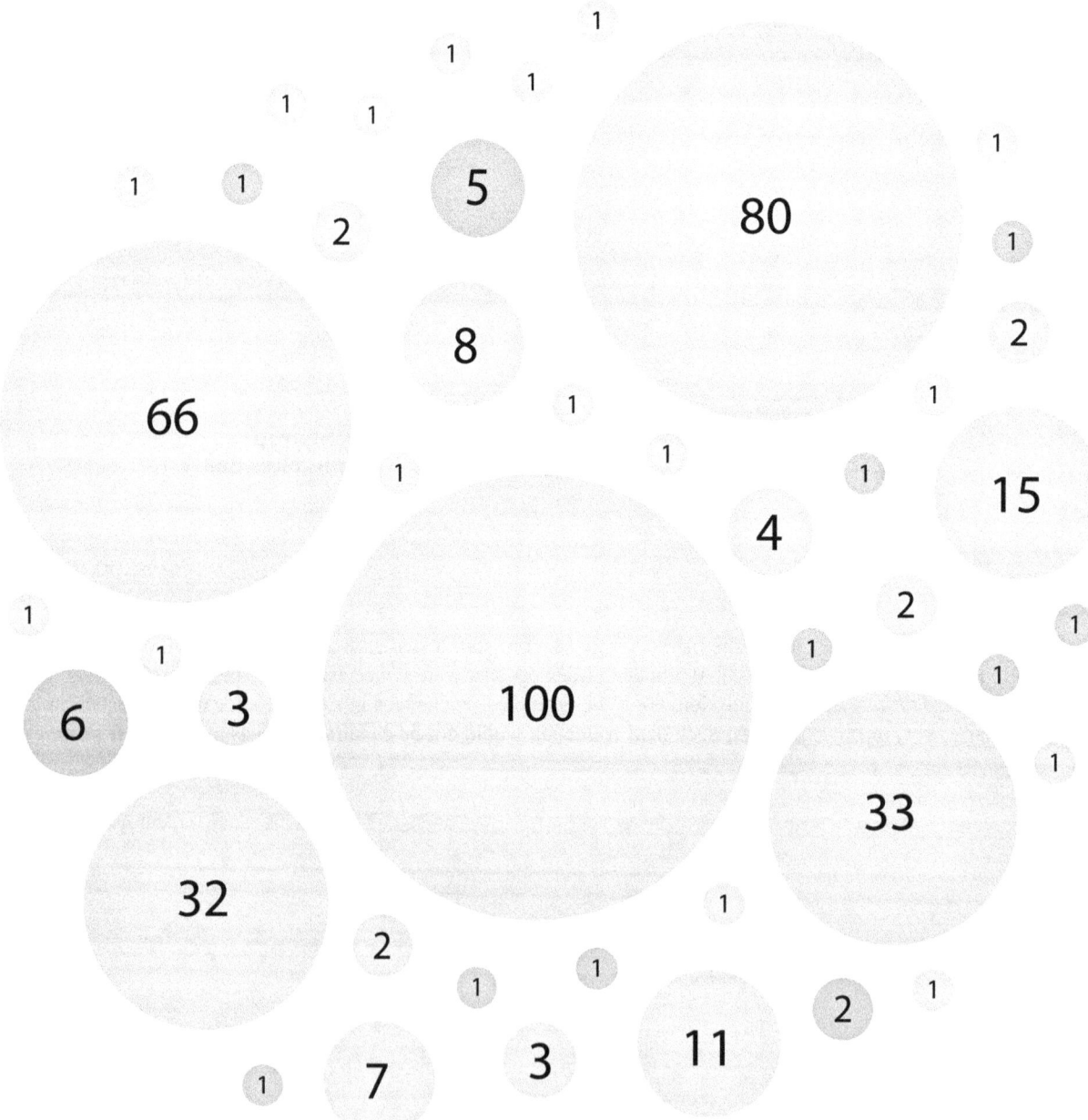

Staff also looked at whether the apps that transmitted information from the device disclosed this fact to users. As noted above, staff adopted an expansive view of what constituted a privacy disclosure that included disclosures or links on the app promotion page, on the developer's website, or in the app itself. Staff also examined the apps' data collection and sharing practices using fairly narrow criteria, searching only for the collection and sharing of specific data points during the short period of time each app was used.

The survey results show that, despite these generous parameters, a significant number of apps transmitted information from the device without disclosing this sharing to users. Indeed, while 59% (235) of the apps transmitted device ID, geolocation, or phone number either to the developer or a third party, only 20% (81) of the apps reviewed provided any privacy disclosures to users. The results also show that most of the data that was transmitted was sent to ad networks, analytics companies, or other third parties. Specifically, 56% (223) of the kids' apps reviewed sent the data to third parties, but only 20% (44) of these apps provided any privacy disclosures. Staff also observed that multiple apps transmitted information immediately upon use, highlighting the need for clear and consistent disclosures to parents prior to an app's download.[27]

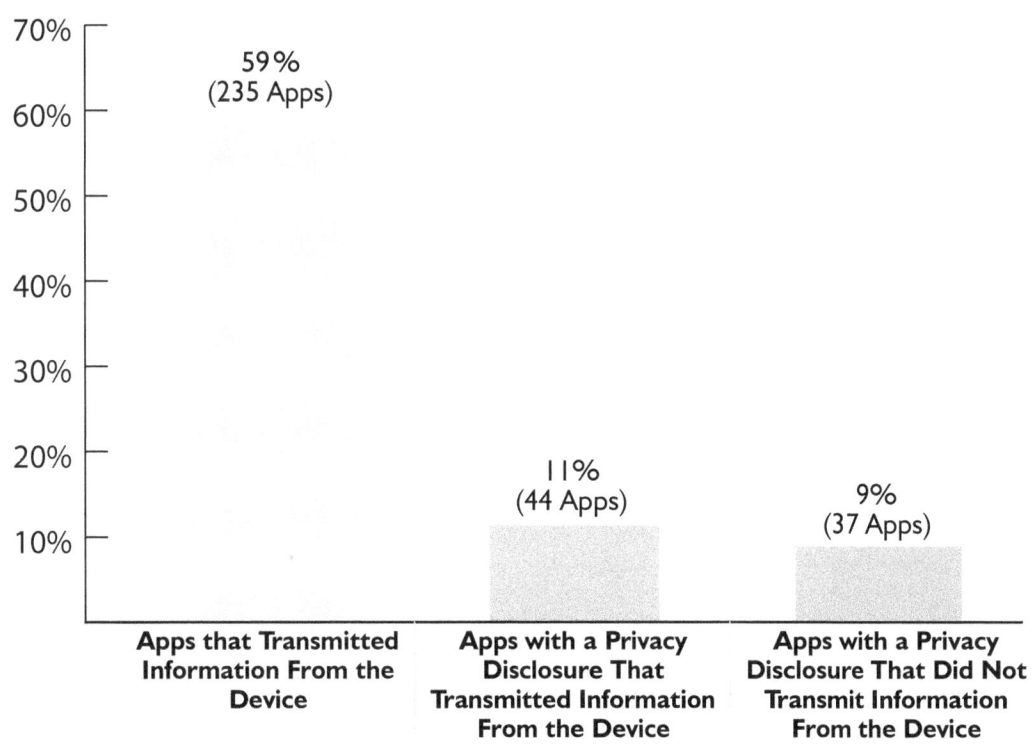

27. One app, for example, transmitted geolocation information to two separate ad networks within the first second of the app's use.

Every device has a unique device ID.

When an app transmits information, it may also send your device ID with that information.

So, when many apps share data with the same company,

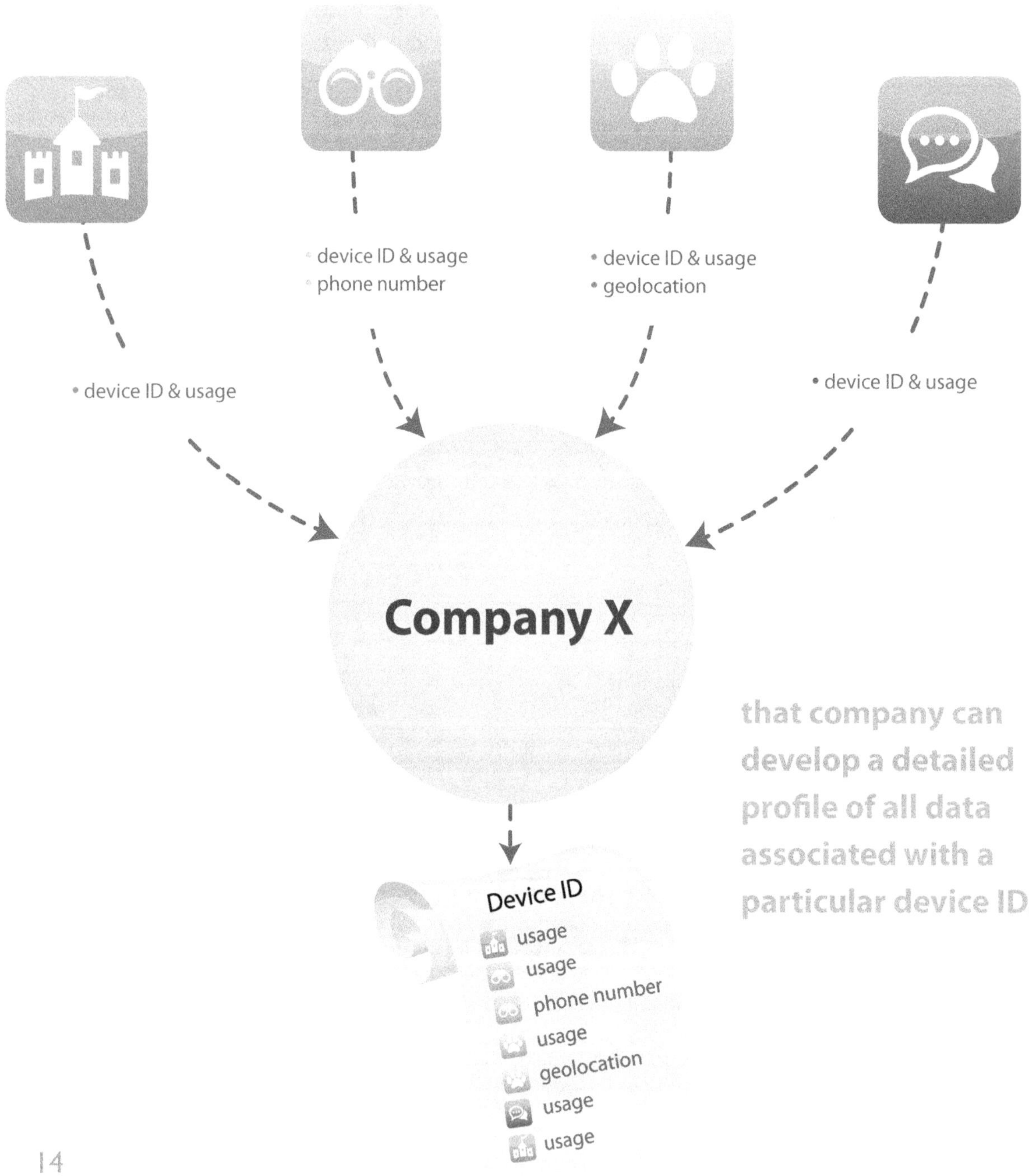

• device ID & usage
• phone number

• device ID & usage
• geolocation

• device ID & usage

• device ID & usage

Company X

that company can develop a detailed profile of all data associated with a particular device ID.

Device ID

usage
usage
phone number
usage
geolocation
usage
usage

The transmission of kids' information to third parties that are invisible and unknown to parents raises concerns about privacy, particularly because the survey results show that a large number of apps are transmitting information to a relatively small number of third parties. Indeed, using the device ID and other information obtained from multiple apps, these third parties could potentially develop detailed profiles of the children using the apps, without a parent's knowledge or consent. Although it is not clear from the survey results whether the information was, in fact, used for this purpose,[28] the frequent transmission of data, coupled with the apps' poor disclosures overall, raises serious questions.[29] In total, staff observed 223 apps transmitting data to one of 30 ad networks, analytics companies, or other third-parties whose precise use or need for the transmitted information was largely undisclosed.

Parents do not interact with these third parties and would be unlikely to see or understand the relevance of these entities' disclosures (assuming that the entities even provided them). Further, parents cannot be expected to search their mobile devices for disclosures from multiple parties to figure out how their complex data collection and sharing practices, taken together, impact their children. The transmission of data to these third parties thus illustrates why parents need clear and accurate privacy information in one easily accessible place. Because each party involved in the mobile app marketplace plays a unique role in

---

28. As noted above, staff examined the disclosures available on the apps' promotion pages and developer websites, and within the apps themselves, for information about the apps' data collection and sharing practices. In general, these disclosures failed to identify the third parties that received the kids' information, let alone how these third parties would use it. For purposes of the survey, staff did not issue requests to the third parties to determine how they used the information. Further, staff was only able to ascertain the identities of the third parties by using a network packet analyzer to capture and analyze the internet traffic associated with each app's use. Certainly, parents attempting to select apps for their children would not have the time or the ability to analyze the internet traffic of each app, or send out requests to third parties, in order to identify who is collecting information from their kids and how they plan to use it.

29. As the Commission has recognized, how certain information is used affects the extent to which privacy concerns are raised. For example, under the FTC's proposed revisions to COPPA, a persistent identifier such as a device ID would be considered "personal information," subject to COPPA's notice and consent requirements, where it can be used to recognize a user over time or across different sites or services, and is not used to support the "internal operations" of the website or online service. The use of device ID for "internal operations," which includes such activities as site maintenance and analysis, performing network communications, authenticating users, maintaining user preferences, serving contextual advertisements, and protecting against fraud and theft, raises lesser privacy concerns and would not be subject to COPPA's notice and consent requirements. *See* Children's Online Privacy Protection Rule; Supplemental Notice of Proposed Rulemaking and Request for Public Comment, 77 Fed. Reg. 151 (Aug. 6, 2012), 46647-46648. *See also* Privacy Report, *supra* note 4, at 36 (noting that the recommendations in the Privacy Report regarding when consumer choice should be provided depends on the context of the interaction between a business and the consumer).

data collection and transmission, each party must play a role in developing effective privacy disclosures for parents.[30]

Disclosures and Practices of Interactive Features of Surveyed Apps

In the first kids' app survey, staff reviewed the app promotion pages of 400 apps to see if they included disclosures regarding whether the apps included advertising, allowed users to make in-app purchases, or linked to any social media.[31] For the follow-up survey, staff again examined the disclosures regarding the interactive features of 400 apps, but also downloaded and reviewed each app to determine how many apps actually contained these interactive features. Staff found that a significant number of the apps contained interactive features, but that many failed to disclose this fact to parents prior to download.

In-App Advertising

The first kids' app survey found that 7% (28) of the 400 apps reviewed disclosed that an app contained advertising. Using the same methodology for the follow-up survey, staff found that 15% (59) of the apps disclosed whether they contained advertising, with 9% (35) stating that they do contain advertising and 6% (24) stating that they do not.

While the increase in the level of disclosure of advertising may seem like a small measure of progress, the gap between the number of apps that actually contain advertising and the number that disclose that fact to parents is troubling. In fact, in contrast to the 9% (35) that disclosed to parents that their apps contained advertising, 58% (230) actually contained ads. And of the 24 apps that stated that they did not contain advertising, ten apps actually *did* contain advertising. Staff also found that multiple apps displayed screenshots of the app on the app promotion pages that showed no advertising when in fact the apps did contain ads, some of which were mature in nature. For example, as shown in the screenshots below, a drawing app's promotion page displayed screenshots of the app containing no ads. However, after downloading and running the app, staff discovered that the app actually contained advertisements for an online dating website.

---

30. The app stores set the technical rules that govern what information apps can and cannot access, and provide app developers with a platform to reach consumers. The app developers collect data from their users, and also may integrate functionality from advertising networks, analytics companies, or other third parties that collect data from users. Each participant thus has unique knowledge about the information they are collecting, or enabling others to collect, from users, and must work together to develop disclosures reflecting these practices.

31. *See* Mobile Apps for Kids, *supra* note 2, at 4.

Screenshot of App from its Promotion Page:

Description

The BEST painting program for kids!
Paint on your picture and have fun!
The BEST painting program for kids!

* Best Free Android Apps and Games for Kids! - AppConsumer
* Simple, interesting and fun for kids and their parents - Android Apps
* If you do happen to have little ones, this is simply a must - BriefMobile

Draw beautiful art with random color Shake the device to clear screen. You can also take

MORE

Visit Developer's Website   Email Developer

App Screenshots

 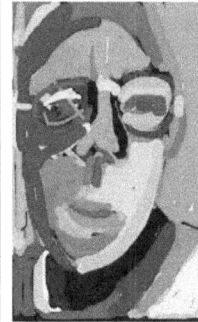 

Screenshot from within the App:

See **1000+** Singles

As noted in the first kids' app report, there are a variety of reasons why parents may have concerns about the presence of advertising within an app that their child will use. During the follow-up survey, staff came across numerous complaints from user feedback ratings on various app promotion pages echoing these concerns, ranging from objections to the content of the advertisements to complaints about the data collection associated with such advertising. While everyone may not share these same concerns, parents should be given the opportunity to make this choice for their children prior to downloading the app.

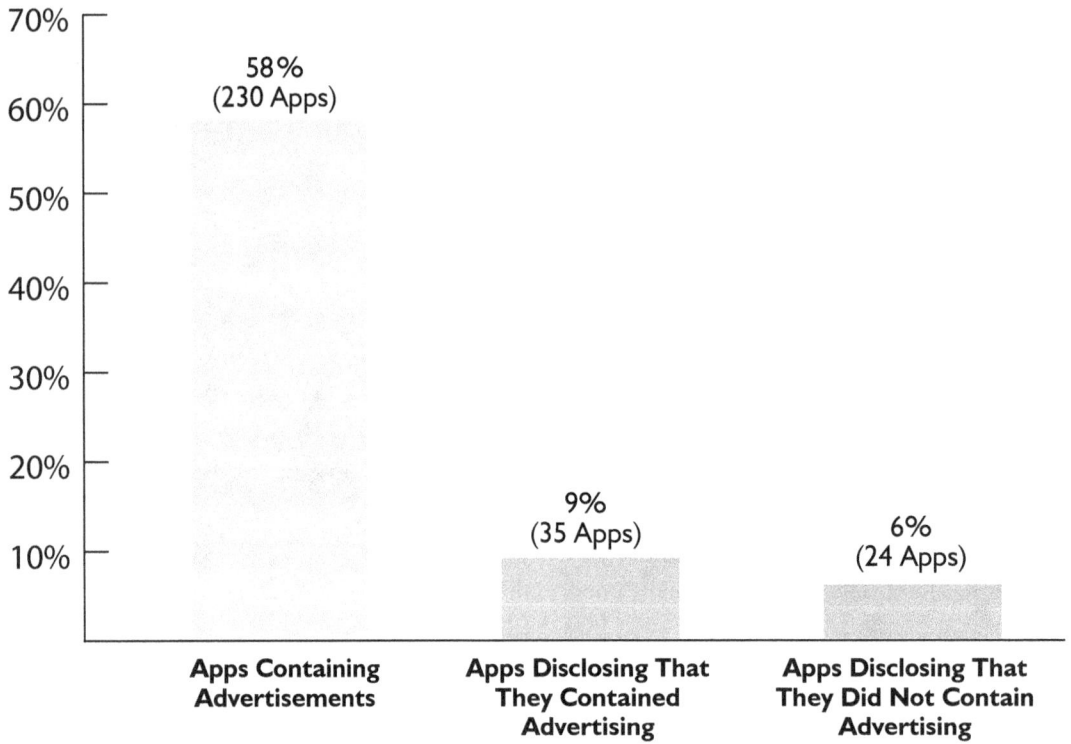

In-App Purchases

The first kids' app report found that approximately 6% (23) of the apps reviewed indicated that the app allowed users to purchase additional content through an in-app purchase mechanism. For example, a kid's game may allow a child to pay $0.99 to purchase a new vehicle to use in a racing game or to advance to a higher level more quickly.

In the follow-up survey, staff found a considerable increase in the number of apps that enabled in-app purchases. In fact, 17% (66) of the apps reviewed allowed users to make purchases within the app, with 3% (6) from the Google Play Store and 30% (60) from Apple's App Store allowing for in-app purchases. By contrast, the first survey found that 11% of the 200 Apple App Store promotion pages and 0.5% of the 200 Android pages reviewed indicated that the app allowed for in-app purchases.[32]

The Apple and Android operating systems provide certain indicators to signal that an app allows users to make purchases within the app. Apple includes a box labeled "Top In-App Purchases" on the promotion page for apps with in-app purchase mechanisms. If users click on the box, they are provided with a list of the items available for purchase within the app. Android discloses in-app purchasing capability under the "Network communication" category

---

32. *See* Mobile Apps for Kids, *supra* note 2, at 13.

on a list of "permissions."[33] Even if parents notice these indicators, however, they may not understand the meaning of the term "in-app purchases" and they may not understand that the apps can be used by children to make frequent and expensive purchases.

Additionally, the indicators do not explain clearly (or provide an easy means for consumers to learn) many important aspects of these purchases, including why such purchases are being offered, whether they may result in recurring charges, whether further parental authorization is required to make them, and what the applicable refund policies are.

The ability of children to purchase items within mobile apps has been the subject of considerable concern among parents, the media, and members of Congress.[34] Parents report that they did not know about such capabilities prior to downloading an app and that their young children were able to make purchases totaling hundreds of dollars without the parents' knowledge. Compounding this confusion is the fact that many of these apps are offered to parents as "free."[35] This problem is likely to grow as the use of mobile devices continues to explode and in-app purchasing becomes an increasingly common business model. Parents should be given the information they need to make informed decisions about whether to allow their children to use apps with these capabilities.

Links to Social Media

Some apps allow users to connect with social media so that users may communicate with their social networking "friends" or "followers" through the app. For example, an app may let users post drawings they create within the app on a social networking site, participate in a chat room, or compare high scores in a game with other users. In the first kids' app survey, 5% (20) of the apps disclosed that they linked to social media – that is, a user could access a social network, and thus share information, through an app. In the current survey, approximately 9% (36) disclosed the links with social media through the app.

While the number of disclosures has doubled since the last survey, the survey found that this number represented only half of the apps that actually linked to social media. Specifically,

---

33. Android requires its apps to declare any potentially sensitive capabilities on a "permissions" screen, which displays just before installing the app. See Appendix II for more information.

34. *See, e.g.,* Matt Brian, *Six Year Old Spends $149.99 On Android In-App Purchase*, TheNextWeb (Apr. 20, 2011) *available at* http://thenextweb.com/google/2011/04/20/six-year-old-spends-149-99-on-android-in-app-purchase/; Mobile Apps for Kids, *supra* note 2, at 13.

35. 84% (56) of the apps reviewed that contained in-app purchasing were offered as "free."

the survey showed that 22% (88) of the apps surveyed actually linked to social media, as compared to the 9% (36) that disclosed that fact.

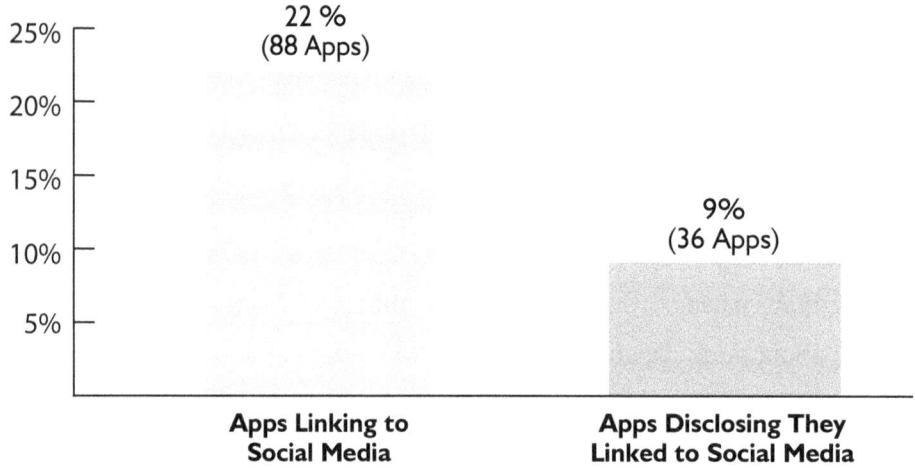

Parents may be concerned about the inclusion of social media in the apps their children use for a variety of reasons. For example, parents may not want their children to communicate with other users who they have never met or to post information about themselves or their whereabouts. Parents may also be worried that their children may post comments, photos, or videos that can damage a reputation or hurt someone's feelings. The presence of social features within an app is therefore highly relevant to parents selecting apps for their children and should be disclosed prior to download.

Apps Containing Social Networking Integration, N=396

Social Network	# of Apps	% of Apps
Facebook	41	10.4%
Google +	36	9.1%
Twitter	29	7.3%
Game Center	6	1.5%
OpenFeint	5	1.3%
YouTube	5	1.3%
HeyZap	2	0.5%
Tumblr	2	0.5%
LinkedIn	1	0.3%
TOTAL	**88**	**22%**

Conclusion

Since FTC staff conducted its first kids' app survey, many stakeholders have called for industry to increase transparency in the mobile marketplace, and many initiatives have been launched in pursuit of that goal. Despite these efforts, staff found little or no improvement in the disclosures made and, worse, a significant discrepancy between the privacy disclosures and the actual practices of the surveyed apps. Without adequate and accurate information about apps they download for their kids, parents cannot make informed choices about their children's privacy and exposure to social networks and other interactive features.

FTC staff has initiated a number of investigations to address the gaps between company practices and disclosures. These discrepancies could constitute violations of COPPA or the FTC Act's prohibition against unfair or deceptive practices. However, enforcement actions alone, while vitally important, are not enough to ensure that the privacy of consumers and their children are protected adequately. Staff calls on everyone involved in the mobile app marketplace – app stores, app developers, and third-parties that interact with the apps – to follow the three key principles laid out in the FTC's Privacy Report: (1) adopting a "privacy-by-design" approach to minimize risks to personal information; (2) providing consumers with simpler and more streamlined choices about relevant data practices; and (3) providing consumers with greater transparency about how data is collected, used, and shared. Of greatest relevance to the findings in this report, industry participants must work together to develop accurate disclosures regarding what data is collected through kids' apps, how it will be used, who it will be shared with, and whether the apps contain interactive features such as advertising, the ability to make in-app purchases, and links to social media.

Appendix I: Methodology

To compare the 2012 survey results to those from the first survey, FTC staff repeated the steps described in the methodology section of last year's report.[36] FTC staff also downloaded and tested each app individually to: examine the presence of additional disclosures within the app; examine the extent to which kids' apps contained interactive features; view what data was transmitted by the app; and identify the recipients of such data.

Review Process: Repeating the Methodology of Last Year's Report

On June 5, 2012, staff used a desktop computer with the Windows 7 operating system to locate and copy the app store promotion pages for 960 mobile applications using the following steps. Staff first searched on the term "kids" in the desktop version of Apple's iTunes App Store and noted that each app had its own nine-digit unique identifier number and its own app store promotion page describing the app. The app store promotion page for each app was viewable by typing in the specific web address within the itunes.apple.com website, which contained the unique app identifier number, into the Internet Explorer browser on the desktop computer. Thus, staff could locate the unique web address for each app store promotion page using the following convention: "http://itunes.apple.com/us/app/id[9-digit-unique-app-id]?mt=8." Staff then used software to visit and copy the browser-viewable app promotion pages for the first 480 apps returned by the "kids" search in the Apple App Store.

Immediately after visiting and copying the first 480 apps returned by the "kids" search in the Apple App Store, staff used the same desktop computer with the Internet Explorer browser to access the desktop version of Google Play, available at https://play.google.com. Staff searched on the term "kids" and noted that each app had its own unique identifier and its own app store promotion page describing the app. Like Apple, the Google Play app promotion page for each app was viewable by typing in the specific web address within the play.google.com website, which contained the unique app identifier, into the browser. Staff could locate the unique web address for each app store promotion page using the following convention: "https://play.google.com/detials?id=[unique-app-id]&feature=search_result." Staff then used software to visit and copy the app promotion pages for the first 480 Google Play apps returned by the "kids" search.

Staff saved each app store promotion page as a .txt file and as an .html file. Staff identified the relevant fields, such as price, developer and number of ratings, found within the

---

36. *See* Mobile Apps for Kids, *supra* note 2, at A1-A7.

compiled app promotion pages and extracted that data into an electronic database. Staff then used a random number generator to select 200 unique numbers and created separate databases containing only the 200 app store promotion pages, pulled from both the iTunes App Store and the Google Play results, which corresponded with the 200 randomly selected numbers. Reviewers were instructed to examine the electronically captured app promotion pages (that had been saved as .html files), and to answer a series of questions about app topic, age range, and disclosures related to their review of the app promotion page. The specific instructions related to this portion of the review are detailed below. Once staff completed the review, two additional reviewers rated the sample, and found almost complete agreement between the first and second review, suggesting that the application of staff's criteria was relatively unambiguous.

Reviewers were also instructed to click on the website address listed on the app promotion pages in the field for "[developer's] website" (and, for the iTunes App Store results, links found in the field for "[App Name] Support"). Staff then saved and reviewed the resulting webpage (the "landing page" of the developer's website), and entered the answers to a series of questions into an electronic form.

In addition, staff expanded on last year's review of the information available prior to download in three ways. First, staff clicked on all links found within the app promotion page descriptions to see if the links resulted in landing pages that provided some form of disclosure. Second, on the developer websites, staff clicked on all links that appeared to lead to disclosures (e.g., staff clicked on all "Privacy Policy" or "Terms of Service" links found on the landing page of the developer website). Finally, staff saved all of the disclosures encountered, making sure to record exactly where the disclosure was found.

In determining what constituted a disclosure, staff relied on the same criteria as in the previous survey. That is, a **privacy disclosure** was considered to be any disclosure labeled expressly as a "privacy" disclosure (such as a "Privacy Policy" or a "Privacy Statement"), or any language that otherwise provided affirmative statements about the app's or developer's information collecting and/or sharing practices. An **interactive feature disclosure** was considered to be any developer-provided statement identifying or explaining the existence of a specific interactive feature (social network, in-app purchasing, or advertising integration) within the app.

Review Process: Downloading and Testing Each App

After repeating the steps from the first survey, staff downloaded and tested all 200 iOS apps and all 200 Android apps.

For a 24-hour period beginning on July 2, 2012, staff downloaded the 200 iOS apps onto two different iPhones and the 200 Android apps onto two different Android devices. Four of the 200 Android apps that were captured on June 5, 2012 were not available for download from Google Play on July 2, 2012. Thus, 196 of the 200 Android apps were downloaded and tested.

Specifically, staff downloaded the 200 iOS apps onto two iPhone 4S smartphones running iOS version 5.1.1, and the 200 Android apps onto two LG Viper 4G LTE smartphones running Android OS version 2.3.7.[37] Each device was assigned a unique static IP address and connected to its own wireless access point. The wireless access points had been configured to connect only to their respective test device, passing internet traffic to and from their test device by way of a bridged connection monitored by a desktop computer. The four desktop computers responsible for bridging the wireless access points to a hardline internet connection had been setup to run software configured to capture the internet traffic passing to and from the test device by way of an IP address capture filter.

Once all the apps had been downloaded onto the test devices, staff played each app once, fully exploring the functionality that would be obvious to a first time user. In interacting with the apps, staff looked for additional disclosures as well as instances of social network, in-app purchase, and advertising integration and recorded their findings in an electronic database, documenting any instances of disclosures and interactive features by taking screenshots. As part of the testing process, staff recorded all internet traffic associated with each app.

To maintain as consistent a test environment as possible, staff also created baseline device configurations from which each app was opened, interacted with, and then closed. Once all of the apps had been tested in this manner, staff reviewed the internet traffic that had been captured. In reviewing the internet traffic, staff looked for transmission of device IDs, phone numbers, and geolocation information by applying filters for the specified values to the captured network traffic.

---

37. At the time when staff began testing the Android apps, 64% of all Android devices were running some version of Android 2.3, aka "Gingerbread." Only 13% of all Android devices were running a more recent version of Android. *See* Roy Alugbue, *Latest Android Distribution Chart Shows Gingerbread 2.3 Is On Most Devices, But Ice Cream Sandwich Is Slowly Creeping Upwards*, Talk Android (July 3, 2012), *available at* http://www.talkandroid.com/120208-latest-android-distribution-chart-shows-gingerbread-2-3-is-on-most-devices-but-ice-cream-sandwich-is-slowly-creeping-upwards/.

App Store Desktop Interface v. Mobile Device Interface, and Other Methodological Concerns

As discussed above, the information reviewed by staff in the first phase of the survey had been copied from the app stores viewable on a desktop computer, nearly a month before the apps were actually downloaded and tested, one time, in the FTC's internet lab. Staff took steps in order to minimize the effects that these methodological constraints had on the results.

There are two differences between the app promotion pages viewable in the desktop versions of the app stores and the app promotion pages viewable in the mobile versions of the app stores. App promotion pages viewed via a mobile device are formatted differently from those viewed from a desktop computer (but the content remains the same), and mobile device users may be taken to different developer websites than desktop users. To reduce the effects of these differences on the results, staff gave equal weight to the content found on the app promotion pages (regardless of formatting) and navigated to the developer websites listed in the app promotion pages from both a mobile device and a desktop computer. Staff found that the differences between the developer websites viewed from a mobile device and those viewed from a desktop computer were negligible.

Because nearly a month had passed between the capturing of the app promotion pages and the downloading of the apps, it was possible that the information found in the app promotion pages could have changed. In order to measure the changes in the app promotion pages between the date of initial capture and the date of download, staff re-captured each app promotion page on July 2, 2012. Staff then compared the two sets of app promotion pages, and found no substantive differences.

Finally, practical considerations prevented staff from fully replicating normal use of the apps in real-life situations, and from completely measuring all observed data transmission. By testing each app one time, in the same location, and on a limited number of devices that were connected to the internet solely over Wi-Fi, staff likely captured minimum levels of data transmission. It is possible that the same apps running on different devices or versions of the operating systems would transmit more or different data points. It is also possible that changes in a device's location, connection to a cellular tower, or other configuration would also trigger additional or different data collection. In addition, staff only reviewed the unencrypted network traffic that had been captured; thus, information sent in an encrypted format would not have been counted.

Appendix II:
Additional Information about Apps Tested

Range of Apps Reviewed

In the follow-up study, like the first one, staff categorized the apps reviewed according to words found within the app descriptions and titles. As in the first survey, staff found a wide range of apps intended to be used by kids and offered at low prices by hundreds of developers.

Types of Apps Offered

Staff first categorized the apps by type, according to words found within the app descriptions and titles. Staff allowed apps to fall into more than one category, *e.g.*, a matching game involving addition and described as educational would fall into the "Game," "Matching," "Math," and "Educational" categories.

Category	% of Apps 2011	% of Apps 2012
Educational	50.5%	52.0%
Game	45.3%	61.0%*
Animal-related	22.5%	28.5.%
Alphabet/Spelling/Words	18.8%	22.8%
Math/Numbers	18.3%	15.3%
Matching	12.8%	9.5%
Memory	14.8%	13.8%
Book/Story	9.5%	9.3%
Coloring	13.0%	18.5%*
Musical	6.5%	11.5%*
Puzzle	7.3%	11.8%*
Learning a language	8.8%	6.0%
Flash Cards	3.5%	5.3%
Photo-related	3.3%	8.8%*
Quiz/Test	2.5%	3.5%
Jokes	1.0%	0.0%*
Other**	11.5%	5.0%*
	n=400	n=400

* Difference between 2011 and 2012 samples is statistically significant.

** Apps labeled "Other" fell into the categories Reference/Guide, Diet/Obesity/Fitness, Medical, Monitoring/Tracking, or Other.

Intended Audience

Staff also categorized apps by intended audience using cues provided in the apps' descriptions. Staff looked for words in app names or descriptions suggesting that the apps were recommend for, or were appropriate for, certain general age groups, such as "infants," "toddlers," "preschoolers," "children," "kids," "parents," and "teachers." Most of the apps – 90% expressly indicated that they were intended to be used by a "kid," child, infant or toddler, or a preschool or elementary school aged child.[38]

General Age Group	% of 2011 Apps		% of 2012 Apps	
Infant/Toddler	7.8%		17.0%*	
Child	48.3%		50.3%	
Kid	70.0%	89.8%	74.0%	88.0%
Preschool	9.0%		14.3%*	
Elementary School	1.5%		0.8%	
Parent	18.8%		20.5%	
Teacher	3.0%	24.3%	2.3%	24.0%
Adult	5.0%		4.0%	
Family	5.3%	5.3%	3.5%	3.5%
Everyone	3.0%	3.0%	9.0%*	9.0%*
No Indication	5.0%	5.0%	3.8%	3.8%
	n=400	n=400	n=400	n=400

* Difference between 2011 and 2012 samples is statistically significant.

Twenty four percent of the 400 apps specified a particular age range or school grade level. For these apps, staff recorded the recommended age ranges, converting any grade levels to ages.[39] Over 80% of the apps that listed an age or grade range listed a range beginning

---

38. Note that 13 apps contained no textual age indication, and 2 others contained no "kid" textual indication (*e.g.,* parent or teacher only). Of these 15 apps, 6 appeared to be intended not for kids. The other 9 appeared to be games that kids would enjoy (*e.g.,* simple strategy games like Checkers and Tetris).

39. Staff converted the grade kindergarten to the age 5, first grade to the age 6, second grade to the age 7, etc.

at four years old or younger. Eighty three percent of the apps that specified an age range specified one ending at 12 years old or younger. The table below lists the number of apps for specified age ranges, displaying the changes from the previous survey in parentheses.

		Maximum recommended age					% of apps with this min. age
		3-4	5-6	7-8	9-12	13+	
Minimum recommended age	0-2	5 (-4)	24 (-3)	16 (+10)	4 (+2)	4 (-7)	61%
	3-4		8 (+7)	9 (+1)	6 (+1)	3 (-6)	30%
	5-6		1 (-3)	3 (+2)	1 (0)	1 (0)	7%
	7-8			0 (-3)	2 (0)		2%
	13					0 (-1)	
% of apps with this max. age		6%	38%	32%	15%	9%	n=87

(+/ - represents difference between 2011 and 2012 samples.)

App Pricing and Popularity

To estimate price level popularity for the Android apps in the survey, staff summed the lower bound of the download range with the upper bound of the download range for each app within a given price level, and then divided each sum by 2. Next, staff divided these midpoint price level sums by the sum of the midpoint download ranges for all 200 Android apps. Staff then used the number of user reviews associated with each app as a second estimate of app popularity for the Android apps. As shown in the table below, 97.50% of the apps were described as free (an increase of 36% from 2011), and these apps constituted over 99% of the total downloads.

Android App Pricing and Popularity

Price	% of Apps	% of Downloads	% of Feedback Ratings
Free	97.50% (+35.80%)	99.33% (-0.10%)	98.55% (+0.91%)
$0.01 to $0.99	0.83% (-13.17%)	0.34% (+0.21%)	0.15% (-0.04%)
$1 to $1.99	0.21% (-8.99%)	0.03% (-0.06%)	0.01% (-0.26%)
$2 to $2.99	0.83% (-2.07%)	0.18% (-0.05%)	0.98% (-0.45%)
$3 to $3.99	0.63% (-0.98%)	0.12% (+0.03%)	0.31% (-0.06%)
$4.00+	0 (-1.60%)	0 (-0.01%)	0 (-0.03%)
	n=480		

Because Google Play and Apple's iTunes App Store both display the number of users that have provided feedback for a particular app, staff used the number of user reviews to estimate price level popularity for the Apple apps. Forty nine percent of the iOS apps were described as free (an increase of over 14%), and these apps were responsible for over 63% of all feedback ratings.

iOS App Pricing and Popularity

Price	% of Apps	% of Ratings
Free	48.96% (+14.16%)	63.44% (-4.85%)
$0.99	30.63% (-13.32%)	26.70% (+4.62%)
$1.99	14.58% (0.00%)	9.57% (+0.42%)
$2.99	4.38% (+0.63%)	0.29% (+0.10%)
$3.99	0.63% (-0.09%)	0.01% (-0.09%)
$4.99	0.83% (-1.05%)	0.01% (-0.18%)
	n=480	

Developers

One hundred and thirty five different developers accounted for the 200 Android kids apps reviewed by staff. The overwhelming majority of these developers were responsible for only one or two apps. Indeed, 74% of the Android developers were each responsible for only one of the 200 Android apps reviewed by staff, and 92% were responsible for no more than two apps.

Number of Android Apps Per Developer

Responsible for # of Apps	1	2	3	4	5	6	7	8
# of Developers	100	24	5	1	1	1	2	1
% of Developers	74%	18%	4%	1%	1%	1%	1%	1%

Similarly, staff encountered 106 different developers in reviewing the 200 iOS apps. Here again, the majority of these developers (72%) were each responsible for only one of the 200 iOS apps reviewed by staff, and 85% were responsible for no more than two apps.

Number of iOS Apps Per Developer

Responsible for # of Apps	1	2	3	4	5	6	7	8	25
# of Developers	76	14	5	2	4	2	0	2	1
% of Developers	72%	13%	5%	2%	4%	2%		2%	1%

In total, 237 developers accounted for the 400 apps that staff reviewed. (Four of these developers were responsible for at least one app in both the 200 Android apps and the 200 iOS apps that staff reviewed.) Staff compared this sample of 237 developers with those developers encountered in the previous survey, and found that 69 developers had at least one app reviewed in both studies. Finding that a sizeable amount of developers were responsible for an app in both surveys, staff then compared the disclosures associated with these developers. Twenty of the 69 developers had a disclosure in both studies, two did not have a disclosure in the 2011 survey but did have a disclosure in the 2012 survey, and three developers had a disclosure in the 2011 survey but not in the 2012 survey. These findings suggest that the lack of disclosures encountered in this survey do not result from a new sample.

Permissions

As discussed in the first kids' app report,[40] Android requires its apps to declare any potentially sensitive capabilities on a "permissions" screen, which displays just before installing the app. Staff found that there was a marked increase in the number of apps declaring permissions that raise concerns for parents. More than 80% of this year's apps contained the ability to access the internet (compared to 62% last year), and more than 13% had the ability to access the user's geolocation (compared to 10.5% last year). Perhaps most telling is the fact that only 9.5% declared the "No special permissions" permission (compared to almost 25% last year), and only 6% declared the "No unsafe permissions" permission (compared to 8% last year).

---

40.   *See* Mobile Apps for Kids, *supra* note 2, at 10-13.

Permission	% of Apps (2011)	% of Apps (2012)
Network communication: full Internet access	62.0%	80.5%
Phone calls: read phone state and identity	19.5%	30.5%
Modify/delete SD card	15.5%	31.0%
Your location	10.5%	13.5%
Fine (GPS) location	6.0%	10.0%
Coarse (network-based) location	5.5%	9.5%
Both Fine and Coarse location	3.3%	6.0%
Hardware controls: take pictures and videos	4.0%	1.0%
Services that cost you money: directly call phone numbers	2.5%	4.0%
Modify global system settings	3.5%	4.0%
Hardware controls: record audio	1.5%	4.0%
Market billing*	0.0%	3.0%
Your personal information: read Browser's history and bookmarks	0.0%	2.0%
Your personal information: read sensitive log data	0.5%	1.0%
Your personal information: read contact data	0.0%	1.0%
No special permissions	24.5%	9.5%
No unsafe permissions	8.0%	6.0%
	n=200	n=200

* The market "Billing" permission indicates that the app contains in-app purchasing features. *See* http://developer.android.com/guide/google/play/billing/billing_integrate.*html* (providing that "if [an] application does not declare the in-app billing permission . . . Google Play will refuse the requests").

Apple does not employ the same permissions model as Android, but Apple does require a notice be provided to the user the first time that an app attempts to acquire the user's location. Apple also provides a system-level indication for the presence of in-app purchase mechanisms. This indication can be found in a box near the description in the app promotion page labeled "Top In-App Purchases." If a user clicks on the "Top In-App Purchases" box, they are provided with a list of the items available for purchase within the app, and the items' respective price.

As explained in the first kids' app report, these system-level disclosures do not provide parents with the information they need to make informed choices about the apps their kids use. Notably, they do not explain clearly (or provide an easy means for consumers to learn) why an app requests the permissions it does, what the app does with such access, how it intends to use the information it obtains, or whether the app shares the information with third parties.[41]

---

41.   *See* Mobile Apps for Kids, *supra* note 2, at 10.

Appendix III:
Separate Statement of Commissioner J. Thomas Rosch
December 7, 2012

Today, I vote in favor of the staff report entitled "Mobile Apps for Kids: Disclosures Still Not Making the Grade." As I have stated before, I strongly support informed consumer choice – requiring clear, complete, and accurate notices about the handling of personal information and allowing consumers to be fully informed about the consequences of the choices they make.[1] Like staff, I am troubled that there has been little or no apparent change by the mobile app industry in the months since staff's prior report highlighted the lack of information available to parents.[2] The mobile apps industry can, and should, do a better job of promoting informed consumer choice.

However, I write separately to reiterate my belief that any enforcement efforts in this area should be based up the "deception" prong, rather than the "unfairness" prong, of Section 5. In particular, any allegation that an industry member has failed to disclose material information about their information collection practices should be framed as either a deceptive representation, a deceptive half-truth, or a deceptive omission. This approach would not only offer more certainty in the privacy area, it would also be in alignment with the promises the Commission has made to Congress in terms of pursuing "unfairness."[3] Even in cases where it could be argued that a deceptive omission would not offer "perfect" certainty, I think that pursuing a case under a deceptive omission theory less uncertain than the unfairness route. Furthermore, in many cases the omission will be in the form of a "half-truth," and the circumstances will be quite clear that additional disclosure was necessary in order to avoid deception.

---

1. Indeed, as I have said previously, I consider the Commission's insistence that such notices be given to be our most significant contribution to consumer protection. *See, e.g.*, J. Thomas Rosch, Comm'r, Fed. Trade Comm'n, *The Evolution of "Privacy Policy" at the Federal Trade Commission: Is It Really Necessary*, Remarks at the Mentor Group (Sept. 14, 2012), *available at* http://www.ftc.gov/speeches/rosch/120914TheMentorGroupBostonParisFrance.pdf; J. Thomas Rosch, Comm'r, Fed. Trade Comm'n, *Information and Privacy: In Search of a Data-Driven Policy*, Remarks at the Technology Policy Institute Aspen Forum (Aug. 22, 2011), *available at* http://www.ftc.gov/speeches/rosch/110822aspeninfospeech.pdf.

2. Fed. Trade Comm'n, FTC Staff Report, *Mobile Apps for Kids: Current Privacy Disclosures are Disappointing* (Feb. 2012), *available at* http://www.ftc.gov/os/2012/02/120216mobile_apps_kids.pdf.

3. *See* Letter from the FTC to Hon. Wendell Ford and Hon. John Danforth, Committee on Commerce, Science and Transportation, United States Senate, Commission Statement of Policy on the Scope of Consumer Unfairness Jurisdiction (Dec. 17, 1980), *reprinted in International Harvester Co.*, 104 F.T.C. 949, 1073 (1984) ("FTC Policy Statement on Unfairness"), *available at* http://www.ftc.gov/bcp/policystmt/ad-unfair.htm; Letter from the FTC to Hon. Bob Packwood and Hon. Bob Kasten, Committee on Commerce, Science and Transportation, United States Senate (Mar. 5, 1982), *reprinted in* FTC Antitrust & Trade Reg. Rep. (BNA) 1055, at 568-570 ("Packwood-Kasten Letter"); 15 U.S.C. § 45(n) (codifying the FTC's modern approach).